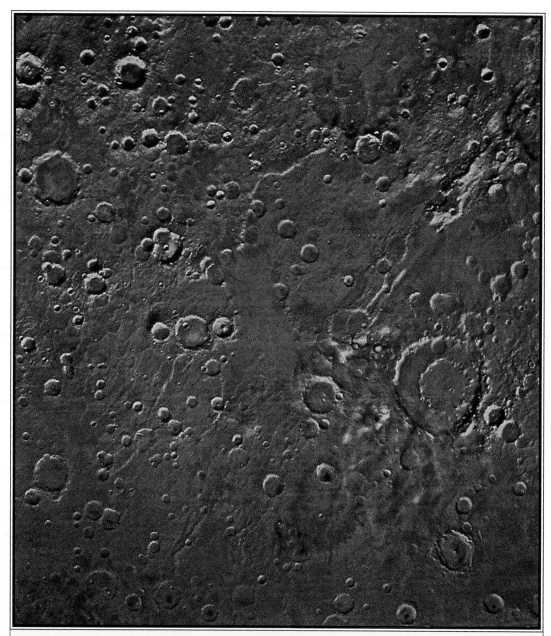

Craters covering the surface of Mars

Mars

Steve Potts

A⁺
Smart Apple Media

COPYRIGHT

Published by Smart Apple Media

1980 Lookout Drive, North Mankato, MN 56003

Designed by Rita Marshall

Copyright © 2002 Smart Apple Media. International copyright reserved in all countries. No part of this book may be reproduced in any form without written permission from the publisher.

Printed in the United States of America

Photographs by Tom Stack & Associates (ESA, JPL, NASA, Inga Spence, TSADO, USGS)

Library of Congress Cataloging-in-Publication Data

Potts, Steve. Mars / by Steve Potts. p. cm. — (Our solar system series)

Includes bibliographical references and index.

ISBN 1-58340-095-8

1. Mars (Planet)—Juvenile literature. [1. Mars (Planet)] I. Title.

QB641 .P68 2001   523.43—dc21   2001020125

First Edition 9 8 7 6 5 4 3 2 1

Mars

Red Planet

One of the brightest objects in the night sky looks like a red ball. It's the red planet, Mars, named after the Roman god of war. One of Earth's nearest neighboring planets, Mars has fascinated people for thousands of years. ☀ When its rotation brings it close to Earth, Mars is only 35 million miles (56 million km) away. At that distance, it is easy to see Mars through a simple **telescope**. Mars has a reddish surface with white polar ice fields, dark markings, and clouds that look

Mars is nicknamed the "red planet"

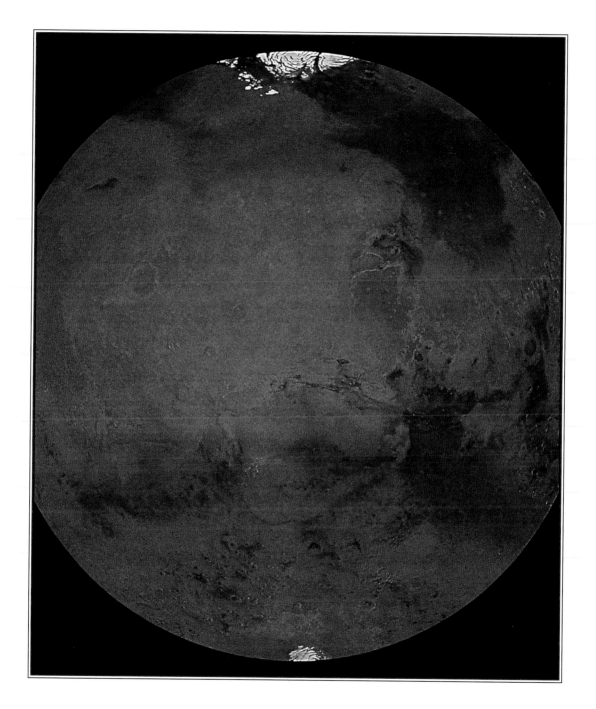

much like Earth's surface. ☀ The first astronomer to view

Mars through a telescope was probably Galileo Galilei. In

1610, Galileo looked at Mars through his primitive telescope.

He saw a surface that reminded him of **One Mars day is only 41 minutes longer than an Earth day; one Mars year equals 687 Earth days.**

Earth. Mars, Galileo found, was a sphere

that revolved around the Sun and seemed

to have changing seasons, like Earth. ☀

Modern-day scientists have found that these seasons last

almost twice as long as Earth's seasons. During summer on

Mars, the ice fields shrink and even seem to disappear.

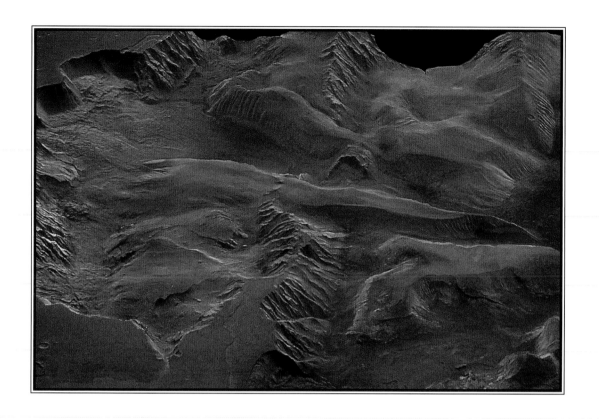

Mysterious Canals

In 1877, Italian astronomer Giovanni Schiaparelli drew a map of Mars that showed what he called "canals." In 1895,

Mars' mysterious "canals"

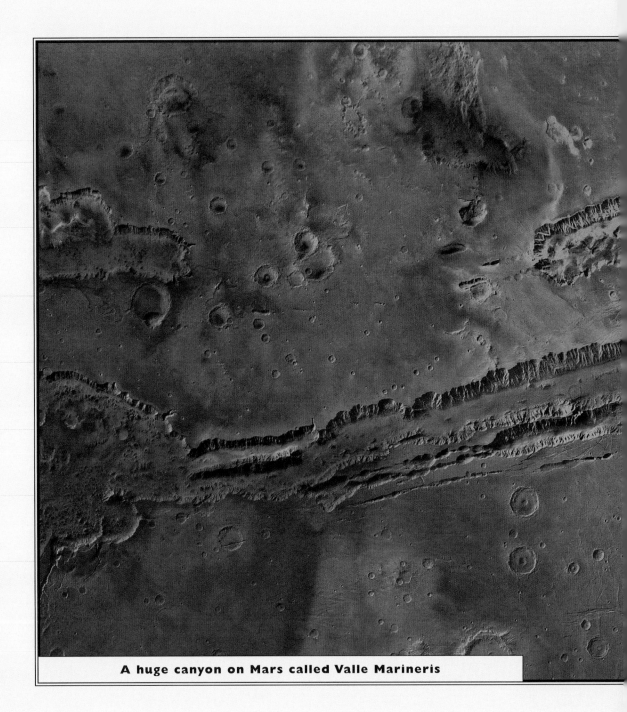

A huge canyon on Mars called Valle Marineris

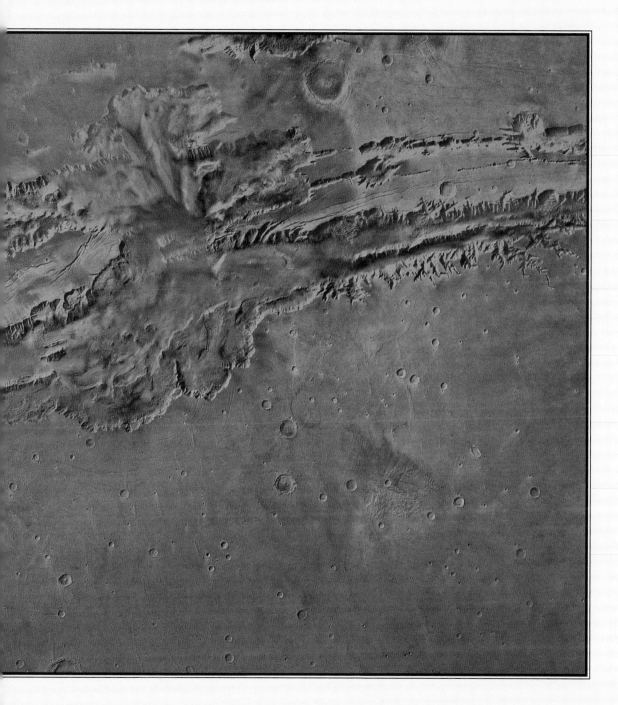

Percival Lowell, an American scientist, said that these canals were built by Martians, people who once lived on Mars. Lowell thought these canals were used to carry water from the polar areas to the dry areas to **irrigate** crops.

In 1960, the Soviet Union became the first nation to try to send spacecraft to Mars.

Now that spacecraft have visited Mars and taken many pictures of its surface, scientists know that these "canals" are actually dry river beds and flash-flood channels. They may have been formed when ice melted below the planet's surface and moved above ground through volcanoes. The water from the melted

ice may have temporarily flooded the landscape before boiling

away. Other scientists think that these worn-away areas are the

remains of a warmer, wetter period in Mars' history. ☀ Mars

Photo of Mars' rocky surface taken by a probe

has two other very interesting features. A volcano called Olympus Mons stands 15 miles (24 km) high, making it the largest in our solar system. A 2,000-mile (3,220 km) canyon called Valle Marineris is 26 times as long and three times as deep as Earth's Grand Canyon.

Martian Atmosphere

Mars, the fourth planet from the Sun, is about half the size of Earth. Its **orbit** around the Sun takes about 687 days, making a Mars year almost twice as long as a year on Earth. A day on Mars is just a little longer than a day on Earth.

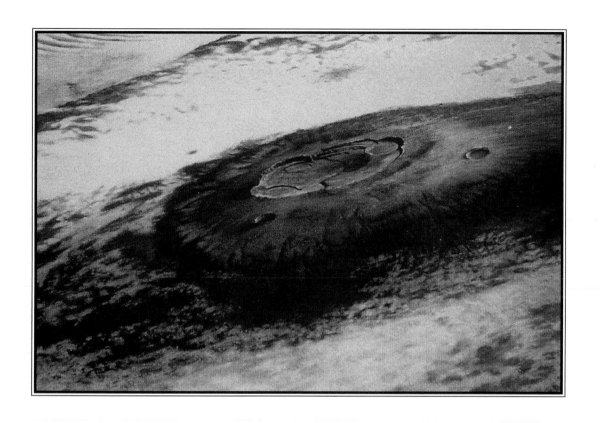

The **atmosphere** on Mars is much thinner than the

atmosphere on Earth. The gases nitrogen and oxygen make up

much of Earth's atmosphere, but Mars has an atmosphere that

The Olympus Mons volcano

is 95 percent carbon dioxide. Humans could not breathe the

air on Mars and survive. Scientists think that this carbon diox-

ide was created when volcanoes erupted on Mars long ago. ☀

The thinner atmosphere means that Mars **Two small**

moons called

can get much colder than Earth can. During **Phobos and**

Deimos orbit

the summer, temperatures on Mars are about **Mars.**

−123° F (−86° C) at night and −23° F (−31° C) in the after-

noon. ☀ In the winter, temperatures on Mars dip to −197° F

(−127° C). At that temperature, clouds made of carbon dioxide

Polar ice cap on the north side of Mars

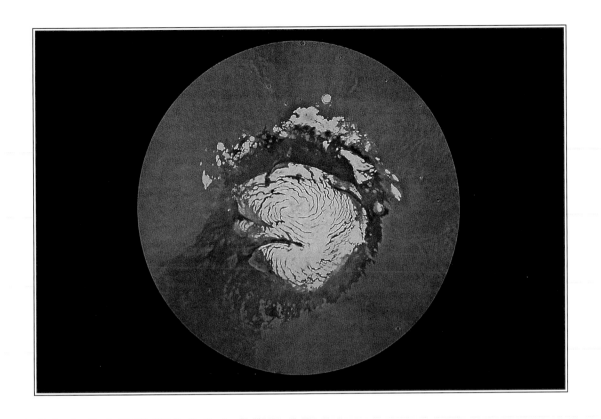

develop. These clouds create "dry ice," the substance that

forms the polar ice caps on Mars.

Exploring Mars

Scientists hope that unmanned probes will reveal more about the red planet. The first probe to Mars was launched by the United States in 1964. Probes are small spacecraft that are sent into space by rockets or carried on a space shuttle. Probes contain complex computer and photographic equipment. The cameras take pictures to send back to Earth by **radio waves**. So far, all of the probes sent to Mars have been unmanned.

In 1976, *Viking 1* and *Viking 2* probes orbited and landed

A radio wave receiver collects data sent back by probes

on Mars and began sending information about the planet's surface back to Earth. In 1997 and 1999, four more American probes were launched. Two of these probes landed on Mars and began a series of scientific experiments that have told us much about the planet. ☀ In the future, humans will probably travel to

Japan plans to send a new probe to orbit Mars in 2004.

Mars—and perhaps even live there. While we probably won't find any Martians, what we do find may reveal many more secrets about the red planet.

Pathfinder, one of the latest probes to explore Mars

Viking 1, the first probe to visit Mars' surface

Futuristic probe collecting soil samples on Mars

Index

Words to Know

atmosphere—the nearly invisible layer of gases that surrounds a planet

irrigate—to use canals and ditches to move water from a wet area to a dry area for farming

orbit—a repeating circular pattern of one object traveling around another

radio waves—energy that travels at the speed of light from a transmitting antenna to a receiving antenna to form a message

telescope—an instrument that uses a glass lens to magnify distant objects

Read More

Bond, Peter. *DK Guide to Space*. New York: DK Publishing, 1999.

Couper, Heather, and Nigel Henbest. *DK Space Encyclopedia*. New York: DK Publishing, 1999.

Ride, Sally, and Tam O'Shaughnessy. *The Mystery of Mars*. New York: Crown, 1999.

Internet Sites

Astronomy.com
http://www.astronomy.com/home.asp

Windows to the Universe
http://windows.engin.umich.edu/

NASA: Just for Kids
http://www.nasa.gov/kids.html

The Nine Planets
http://seds.lpl.arizona.edu/nineplanets/nineplanets/

INFORMATION